Houghton Mifflin California Math

Unit 5

Subtract 2-Digit Numbers

BIG IDEAS!

- You can use drawings, a hundred chart, or you can line up digits to subtract two-digit numbers.
- You can check a subtraction answer by adding.

Songs and Games

Math Music Track 5
How Many Will Be Left?

eGames
www.eduplace.com/camap/

Literature

Read Aloud Big Book

- The Mystery of the Missing Eggs

Math Readers

ISBN-13: 978-0-618-95738-5
456789- WC - 16 15 14 13 12 11 10 09

Get Ready **Game**

Subtract from 99

How to Play

1. Write the number 99 at the top of a piece of paper.
2. Take turns drawing a number card from the bag.
3. Subtract the number from 99. Write the difference on the paper.
4. Take turns drawing new number cards. Subtract each new number from the last difference on your paper.

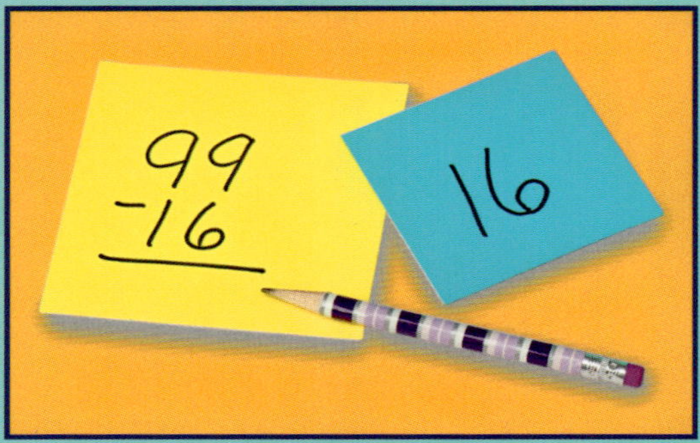

5. The player with the lesser difference after four turns wins.

What You Need

2 players

Learning Tool 31

KEY NS 2.2 Find the sum or difference of two whole numbers up to three digits long.
Also **NS 2.0, MR 1.2, MR 2.2**

Education Place
Visit **www.eduplace.com/camap/** for eGames and Brain Teasers.

Math at Home

Dear Family,

My class is starting Unit 5, **Subtract 2-Digit Numbers.** I will be learning different strategies to help me subtract 2-digit numbers, such as estimating differences and regrouping tens. You can help me learn these vocabulary words, and we can do the Math Activity together.

From,

____

Vocabulary

difference The result of subtraction

$15 - 10 = 5$ (↑ difference)

$$\begin{array}{r} 15 \\ -10 \\ \hline 5 \end{array}$$

difference → 5

estimate A rounded answer that is close to an exact answer

29 —round to nearest ten→ 30

−21 —round to nearest ten→ −20

estimate → 10

regroup In subtraction, trade 1 ten for 10 ones.

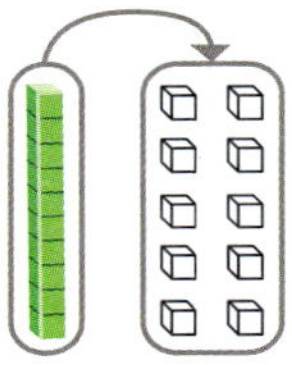

Education Place

Visit **www.eduplace.com/camaf/** for

- eGames and Brain Teasers
- Math at Home in other languages

Family Math Activity

Make up a subtraction number sentence using 2 two-digit numbers, such as 52−36. Have your child tell you a story to go with the number sentence. Then have your child tell if he or she needed to regroup to subtract and why or why not. Add to check the answer.

Literature

These books link to the math in this unit. Look for them at the library.

- **Shark Swimathon**
  by Stuart J. Murphy
  Illustrated by Lynne Cravath
  (HarperCollins, 2001)
- **Lights Out!**
  by Lucille Recht Penner
  Illustrated by Jerry Smath
- **Panda Math**
  by Ann Whitehead Nagda

Matemáticas en casa

Estimada familia:

Mi clase está comenzando la Unidad 5, **Resta de números de 2 dígitos**. Aprenderé diferentes estrategias para restar números de 2 dígitos, como estimar diferencias y reagrupar decenas. Me pueden ayudar a aprender estas palabras de vocabulario y podemos hacer juntos la Actividad de matemáticas para la familia.

De:

____

Vocabulario

diferencia El resultado de la resta.

$15 - 10 = 5$ (↑ diferencia)

$$\begin{array}{r} 15 \\ -10 \\ \hline 5 \end{array}$$

diferencia → 5

estimación Respuesta aproximada que es cercana a una respuesta exacta.

29	redondear a la decena más cercana →	30
−21	redondear a la decena más cercana →	−20
	estimación →	10

reagrupar En la resta, cambiar 1 decena por 10 unidades.

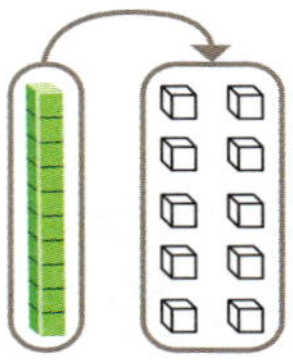

Education Place

Visite **www.eduplace.com/camaf/** para
- Juegos en línea y acertijos
- Matemáticas en casa en otros idiomas

Actividad de matemáticas para la familia

Invente un enunciado numérico de resta usando 2 números de dos dígitos, como 52−36. Pida a su niño que le cuente un cuento que vaya con el enunciado numérico. Luego pregúntele si tuvo que reagrupar para restar y por qué. Sume para comprobar la respuesta.

Literatura

Estos libros hablan sobre las matemáticas de esta unidad. Búscalos en la biblioteca.

- **Katy no tiene bolsa**
  por Emmy Paine
  (Rebound by Sagebrush, 2000)
- **Primeros números**
  por Jo Litchfield

Imagine you go to a sale. You see these 10 dinosaurs for sale. Circle the ones you would buy. How many would be left?

_____ – _____ = _____

_____ would be left

KEY NS 2.2 Find the sum or difference of two whole numbers up to three digits long.

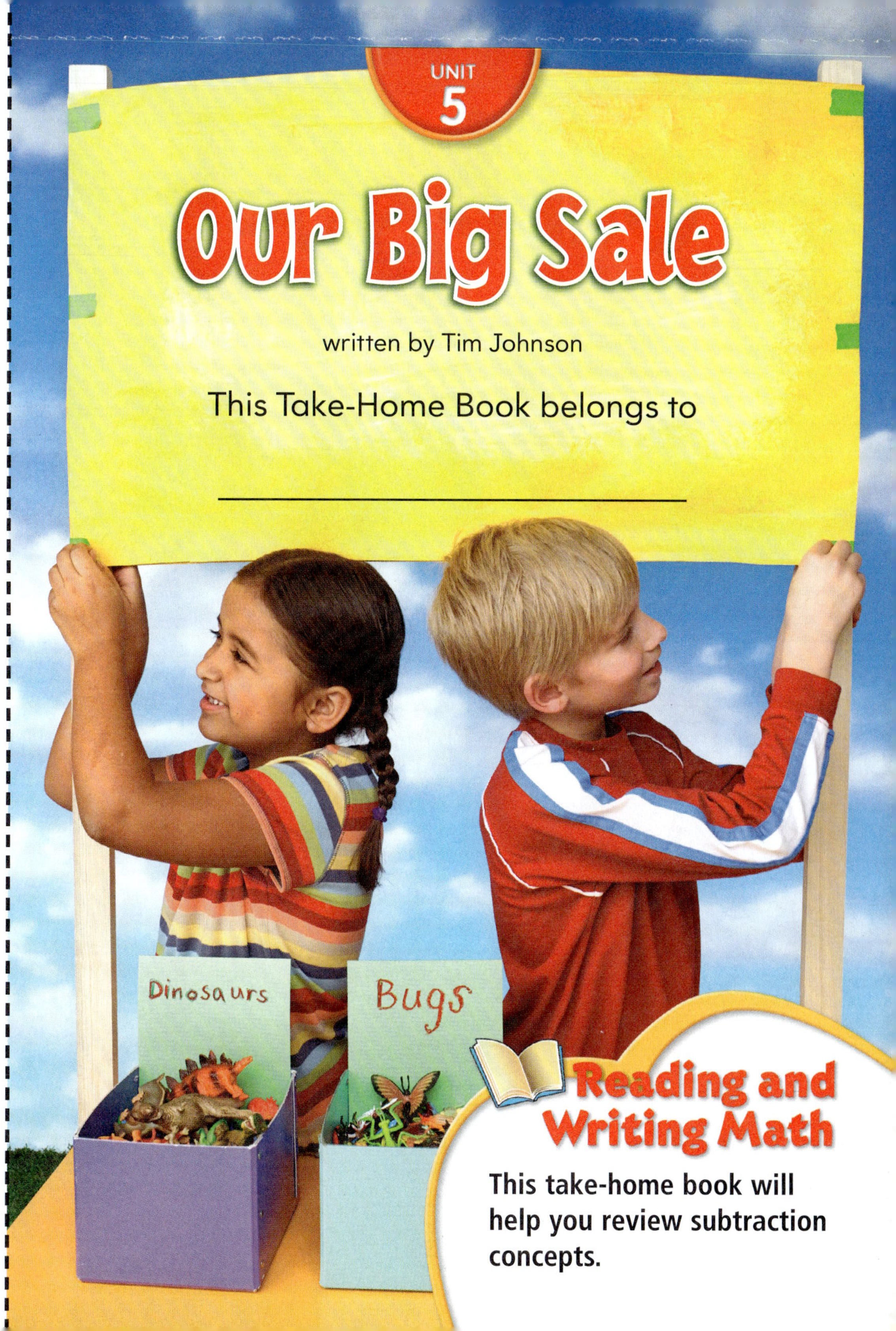

UNIT 5

Our Big Sale

written by Tim Johnson

This Take-Home Book belongs to

____

Reading and Writing Math

This take-home book will help you review subtraction concepts.

Kim and Daren set up a store.
They want to sell their collections of bugs and dinosaurs.
"Come to our big sale," they call.

Kim places the rest of her bug collection on the table.
José wants to buy 9 of the bugs.
How many will be left to sell?

_____ – _____ = _____

_____ left to sell

Along comes Ben. He looks at the dinosaurs Daren still has for sale. How many are in the collection? If Ben buys 5, how many will be left to sell?

_____ – _____ = _____

_____ left to sell

Daren has 6 T-Rex dinosaur toys. Bill comes to buy 2. How many will be left to sell?

_____ – _____ = _____

_____ left to sell

Tina likes Kim's ladybug collection.

If Tina buys 7 of Kim's ladybugs,
how many will Kim have left?

_____ – _____ = _____

_____ left

Chapter 11

Explore Subtracting Tens and Ones

Vocabulary

Here are some vocabulary words you will learn in the chapter.

subtract Take away part of a group

$$7 - 3 = 4$$

↑ minus sign

difference The result in a subtraction problem

$$7 - 3 = 4$$

↑ difference

subtract tens You can take away groups of ten to solve a subtraction problem.

$$90 - 30 = 60$$

See English-Spanish Glossary pages 573–589.

KEY NS 2.2 Find the sum or difference of two whole numbers up to three digits long.

Education Place
Visit www.eduplace.com/camap/ for the eGlossary and eGames.

Name ______________________________

Check What You Know

Write the number sentence.

1. Ted makes 13 paper airplanes. Alan makes 7 paper airplanes. How many more paper airplanes does Ted make?

   ______ − ______ = ______ more paper airplanes

2. Jeff makes 9 paper boats. Dora makes 5 paper boats. How many more paper boats does Jeff make?

   ______ − ______ = ______ more paper boats

3. Clara makes 18 flags. She tapes 9 flags to the window. How many flags are not taped to the window?

   ______ − ______ = ______ flags

4. Pearl draws 12 trees on her picture. She cuts 8 of them out. How many trees are not cut out of the picture?

   ______ − ______ = ______ trees

5. Kayla makes 15 flowers. Steve makes 6 flowers. How many more flowers does Kayla make?

   ______ − ______ = ______ more

Use this page to review important skills needed for this chapter.

Name ____________________

Chapter 11 Lesson 1

Subtract Tens

Hands On

Objective
Use tens rods and basic facts to subtract tens.

Explore

When you subtract tens, you can use a subtraction fact to help.

Patty has 50 pennies. She wants to know how many she will have left if she gives her brother 30 pennies.

Step 1

Use five tens rods to show 50 pennies. Move 3 tens to the side. How many tens rods do you have left? ____

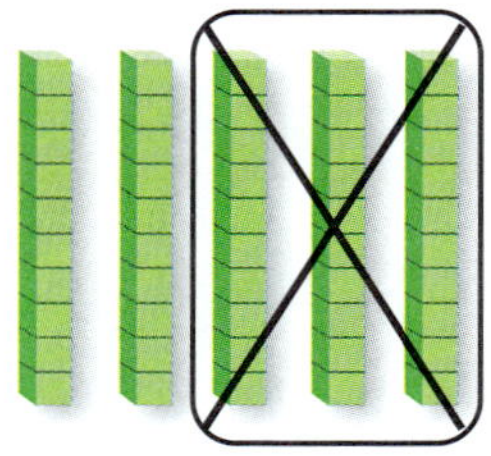

Step 2

What number do two tens rods stand for? ____
Write a subtraction fact that helps solve the problem.

5 – 3 = 2

So, 50 – 30 = 20.

Patty will have 20 pennies left.

1. Use [tens rod] and quick pictures. Use a basic fact to help you.

Think!
A subtraction fact can help me subtract the tens.

Take some [tens rods].

Draw your tens. Subtract some tens by crossing out.

Record.

____ tens – ____ tens = ____ tens

____ – ____ = ____

2. **Math Talk** How does 7 – 2 help you find 70 – 20?

KEY NS 2.2 Find the sum or difference of two whole numbers up to three digits long.
Also **NS 2.0, MR 2.0, MR 3.0**

MR 1.2 Use tools, such as manipulatives or sketches, to model problems.

Extend

Remember!
Think about subtraction facts to help subtract tens.

Use .
Complete each subtraction sentence.
Use a basic fact to help.

3. Find 80 − 40.

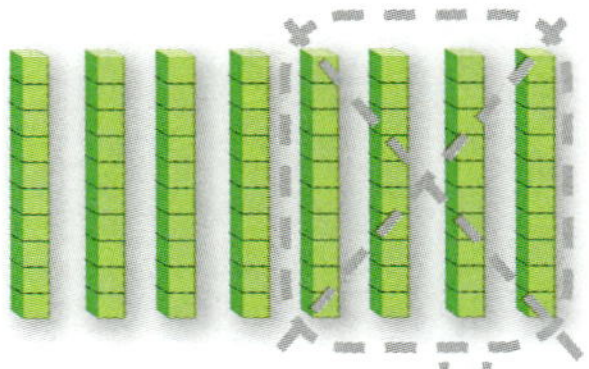

8 − 4 = 4

8 tens − 4 tens = 4 tens

80 − 40 = 40

4. Find 60 − 30.

6 − 3 = ____

6 tens − 3 tens = ____ tens

60 − 30 = ____

Complete the subtraction sentences.
Use to help.

5. 5 tens − 1 ten = ____ tens

____ − ____ = ____

6. 7 tens − 2 tens = ____ tens

____ − ____ = ____

7. 9 tens − 3 tens = ____ tens

____ − ____ = ____

8. 4 tens − 2 tens = ____ tens

____ − ____ = ____

9. 5 tens − 2 tens = ____ tens

____ − ____ = ____

10. 7 tens − 3 tens = ____ tens

____ − ____ = ____

11. 9 tens − 7 tens = ____ tens

____ − ____ = ____

12. 8 tens − 1 ten = ____ tens

____ − ____ = ____

13. How can knowing how to subtract tens help you subtract 27 from 37? Write the subtraction problem.

____

At Home Give your child up to 9 dimes. Create subtraction sentences, such as 8 dimes − 3 dimes = 5 dimes.

En casa Dé a su niño hasta 9 monedas de diez centavos. Creen enunciados de resta, como 8 monedas de diez centavos − 3 monedas de diez centavos = 5 monedas de diez centavos.

Name ____________________

Chapter 11 Lesson 2

Subtract 2-Digit Numbers

Hands On

Objective
Subtract 2-digit numbers using models and drawings.

Learn

Jim has 45 toy dinosaurs.
He gives 13 dinosaurs away.
How many dinosaurs does Jim have left?

Step 1

Show 45 with tens and ones.

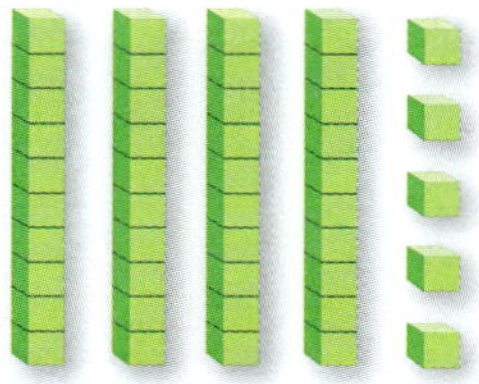

Model the problem with a quick picture.

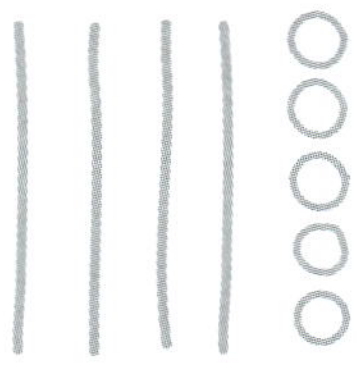

Step 2

Take away 13. Cross out 13 blocks.

Cross out 1 ten and 3 ones in your drawing.

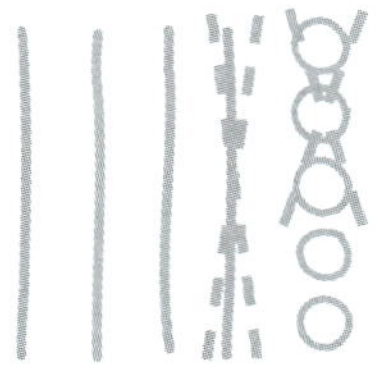

Step 3

Count the tens and ones that are left.

3 tens

2 ones

45 − 13 = 32

Jim has 32 dinosaurs left.

Guided Practice

Model with tens and ones or draw a picture. Subtract.

1. 37 − 15 = ____

Think!
I need to subtract 1 ten and 5 ones.

2. 48 − 31 = ____

3. 64 − 23 = ____

4. 25 − 12 = ____

5. **Math Talk** How does showing a number as tens and ones help you subtract?

▶ Practice

Model or draw to subtract.
Find each difference.

Remember!
Find how many ones and tens you have left.

6. 76 − 45 = 31

7. 39 − 15 = ____

8. 84 − 41 = ____

9. 28 − 14 = ____

10. 36 − 24 = ____

11. 57 − 32 = ____

12. 48 − 25 = ____

13. 68 − 24 = ____

14. 55 − 33 = ____

15. 100 − 70 = ____

Problem Solving: Number Sense

Read the clues. Write the number.

16. Maria has 42 dinosaur stickers. When Maria and Jordan add their stickers together, they have 75 stickers in all. How many stickers does Jordan have?

____ stickers

Draw or write to explain.

At Home Give your child 5 dimes and 9 pennies. Ask him or her to name the amount and show how to subtract 14¢.

En casa Dé a su niño 5 monedas de diez centavos y 9 monedas de un centavo. Pídale que nombre la cantidad y muestre cómo restar 14¢.

Name ______________________

Chapter 11 Lesson 3

Subtract 2-Digit Numbers With Regrouping

Hands On

Objective
Subtract with and without regrouping.

Learn

Joon has 31 butterfly pictures.
He gives 7 pictures to his sister.
How many pictures does Joon have left?

If there are not enough ones to subtract, regroup 1 ten as 10 ones.
Model the subtraction with tens and ones.

Step 1

Show 3 tens and 1 one.

Step 2

Regroup 1 ten as 10 ones.

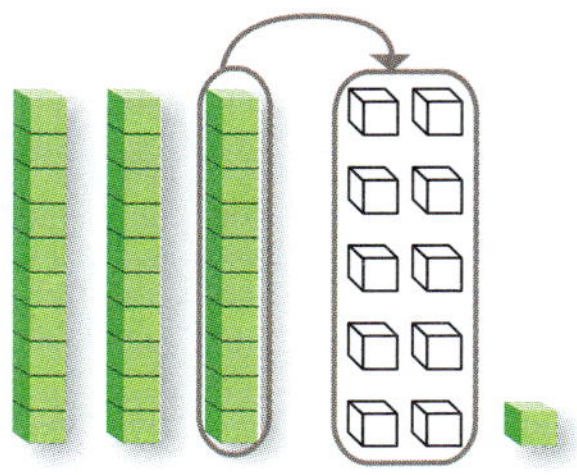

Step 3

Take away 7 ones.

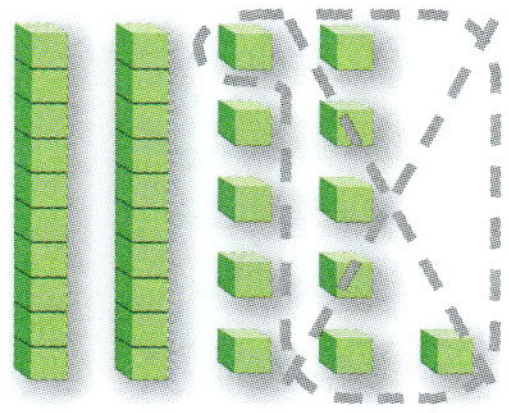

How many are left?

__2__ tens __4__ ones = __24__

Guided Practice

Subtract.
Use tens and ones to help.

Think!
Do I have enough ones to subtract?

	Show the greater number.	Subtract the smaller number.	Do you need to regroup?	How many tens and ones are left?	What is the difference?
1. 42 − 7	42	7	yes no	____ tens ____ ones	____

2. **Math Talk** How do you know when you need to regroup?

Practice

Subtract.
Use [tens block] and [ones block] to help.

Remember!
Regroup if you do not have enough ones to subtract.

3. 33 − 8
2 tens 5 ones
25

4. 56 − 37
____ tens ____ ones
____

5. 37 − 7
____ tens ____ ones
____

6. 65 − 13
____ tens ____ ones
____

7. 24 − 17
____ tens ____ ones
____

8. 72 − 9
____ tens ____ ones
____

9. 98 − 35
____ tens ____ ones
____

10. 100 − 6
____ tens ____ ones
____

Problem Solving: Reasoning

11. Rita has 25¢.
She gives her friend 8¢.
How much money does Rita have left?

Rita has ____¢ left.

Draw or write to explain.

At Home Have your child explain how he or she knew whether to regroup in Exercises 4 and 5.

En casa Pida a su niño que explique cómo supo si debía reagrupar en los Ejercicios 4 y 5.

Name ___________________________

Chapter 11 Lesson 4

Make a Table

Problem Solving Strategy

Objective
Make a table to solve problems.

Learn

Kelly collects things she finds around the house. The pictures show what she finds. Count the objects and complete the table.

How many more buttons than paper clips does Kelly have?

Understand

What do you know?
Look for the information in the table.

- Kelly has 12 paper clips.
- Kelly has 26 buttons.

Kelly's Collection

Object	Number
buttons	
paper clips	
keys	

Plan

Will you add or subtract?

subtract

Think!
I need to compare two amounts, so I will subtract.

Solve

Subtract to solve.

26 − 12 = 14

Look Back

Did you answer the question?
Does your answer make sense?

KEY NS 2.2 Find the sum or difference of two whole numbers up to three digits long.
Also NS 2.0, SDAP 1.1, SDAP 1.4, MR 2.0, MR 2.1, MR 2.2, AF 1.2

AF 1.3 Solve addition and subtraction problems by using data from simple charts, picture graphs, and number sentences.

Guided Problem Solving

1. Joshua collects buttons. The picture shows his collection. Complete the table. How many orange and green buttons does Joshua have?

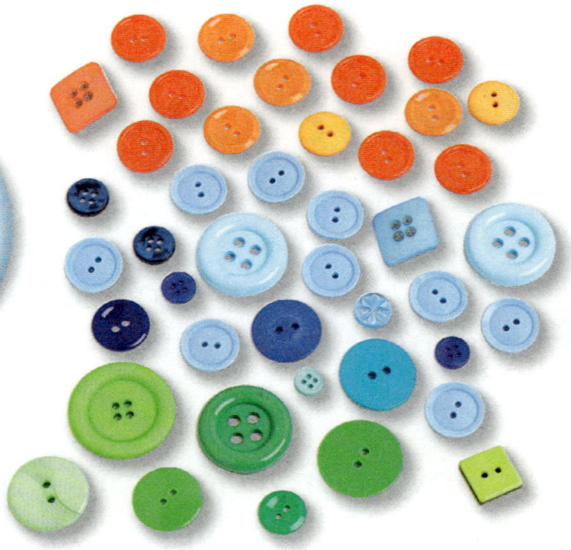

Remember!
Understand
Plan
Solve
Look Back

What do you know?

Joshua has _____ orange buttons

and _____ green buttons.

Will you add or subtract?

Think!
I need to find a total.

____

Write a number sentence.

_____ + _____ = _____

Joshua has _____ orange and green buttons.

Joshua's Collection

Color	Number
orange	
blue	
green	

2. **Math Talk** How did making a table help you solve?

Problem Solving Practice

Use the table to solve.

Draw or write to explain.

3. How many more blue buttons than orange buttons does Joshua have?

_____ more blue buttons

4. How many more blue buttons than green buttons does Joshua have?

_____ more blue buttons

At Home Have your child create an addition problem using the data in the table.

En casa Pida a su niño que haga un problema de suma usando los datos de la tabla.

Name ____________________

Key Standards Review

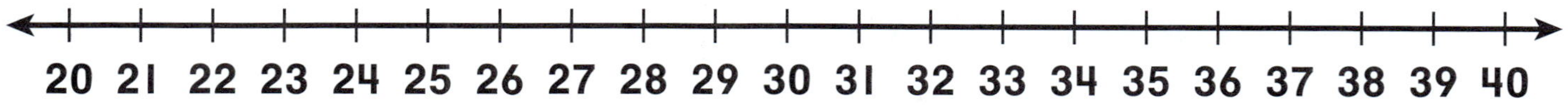

Round each addend to the nearest ten.
Estimate the sum.

1. 32 + 38

_____ + _____ = _____

2. 26 + 33

_____ + _____ = _____

3. 39 + 21

_____ + _____ = _____

4. 35 + 36

_____ + _____ = _____

Add.

5. 25¢ + 56¢ = _____

6. 68¢ + 14¢ = _____

7. 89¢ + 5¢ = _____

8. 36¢ + 48¢ = _____

Challenge Number Sense

Choose the correct number from the box.

You say this number when you start at zero and count by 5s. You say this number when you start at zero and count by 2s. This number is double another number in the box.

What is the number?

_____

15		32
	12	8
25		20
	10	

Science Link

Native American Flutes

Flutes have been part of Native American music for hundreds of years. Flutes are made of many kinds of wood, such as redwood. Music is made by covering holes, blowing into the flute, and creating a vibration.

Write a number sentence. Solve. Draw or write to explain.

1. Mr. Roberts sells 13 flutes in May. He sells 25 flutes in June. How many more flutes does he sell in June than in May?

   _____ − _____ = _____

2. 59 people play the flute. 33 people play the guitar. How many more people play the flute?

   _____ − _____ = _____

3. Mr. Roberts cleans 10 flutes in the morning. He cleans 17 in the afternoon. How many more flutes does he clean in the afternoon?

   _____ − _____ = _____

KEY NS 2.2 Find the sum or difference of two whole numbers up to three digits long.
Also **NS 2.0**
Science PS 1.g

Name ____________________

Concepts and Skills

Complete the subtraction sentence. Use a quick picture or to help. KEY NS 2.2

1. 8 tens − 5 tens = _____ tens

_____ − _____ = _____

Model and draw to subtract. Find the difference. KEY NS 2.2

2. 38 − 15 = _____

Subtract. Use and to help. KEY NS 2.2

3. 52 − 8

_____ tens _____ ones

_____

4. 35 − 29

_____ tens _____ ones

_____

Problem Solving

KEY NS 5.1, NS 5.0, MR 2.0

5. Tina has 63¢. She gives her sister Lucy 24¢. How much money does Tina have left?

_____

Spiral Review and Test Practice

1. Which number is 10 more than 60?

50 ○ 60 ○ 70 ○ 6010 ○

KEY NS 2.0, NS 2.3 Page 189

2. Find the value of the sets of coins.
Which symbol makes this true?

> ○ = ○ < ○ – ○

KEY NS 5.1, NS 5.0, NS 1.3 Page 111

3. What is the sum?

$$\begin{array}{r} 24 \\ +\ 5 \\ \hline \end{array}$$

28 ○ 29 ○ 30 ○ 245 ○

KEY NS 2.2 Page 151

4. What is the sum?

$$\begin{array}{r} 29 \\ +67 \\ \hline \end{array}$$

86 ○ 89 ○ 96 ○ 98 ○

KEY NS 2.2, NS 2.0 Page 175

Education Place
Visit www.eduplace.com/camap/ for Test-Taking Tips and Extra Practice.

Chapter 12

Use Place Value to Subtract

Vocabulary

Here are some vocabulary words you will learn in the chapter.

hundred chart
You can use a hundred chart to subtract.

96 − 40 = 56

1	2	3	4	5	6	7	8	9	10
11	12	13	14	15	16	17	18	19	20
21	22	23	24	25	26	27	28	29	30
31	32	33	34	35	36	37	38	39	40
41	42	43	44	45	46	47	48	49	50
51	52	53	54	55	56	57	58	59	60
61	62	63	64	65	66	67	68	69	70
71	72	73	74	75	76	77	78	79	80
81	82	83	84	85	86	87	88	89	90
91	92	93	94	95	96	97	98	99	100

regroup Trade 1 ten for 10 ones

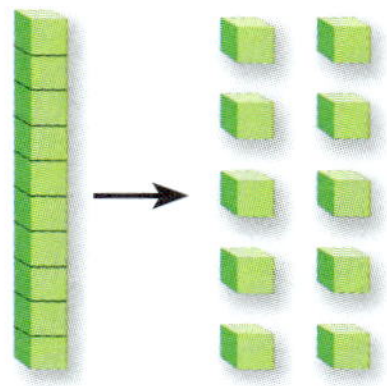

1 ten is the same as 10 ones.

See English-Spanish Glossary pages 573–589.

KEY NS 2.2 Find the sum or difference of two whole numbers up to three digits long.

Education Place
Visit www.eduplace.com/camap/ for the eGlossary and eGames.

Name ______________________________

Check What You Know

Solve. Use the table.

Collections	
stamps	56
rocks	40
buttons	20
keys	14

1. How many stamps are in the collection?

   _____ stamps

   _____ tens _____ ones

2. How many more rocks than buttons are there?

   _____ − _____ = _____ more rocks

3. If 10 buttons were taken out of the button box, how many buttons would be left?

   _____ − _____ = _____ buttons

4. How many more stamps than rocks are in the collection?

   _____ − _____ = _____ more stamps

5. What is the difference between the number of stamps and the number of keys?

   _____ − _____ = _____

Use this page to review important skills needed for this chapter.

Name ______________________________

Chapter 12 Lesson 1

Subtract Using a Hundred Chart

Hands On

Objective
Use a hundred chart and count back by tens to subtract.

Learn

Use the hundred chart.
Find 57 − 30.

Step 1

Find 57 on the hundred chart.

Step 2

Move up 3 rows to subtract 30.

(57), 47, 37, 27

57 − 30 = 27

1	2	3	4	5	6	7	8	9	10
11	12	13	14	15	16	17	18	19	20
21	22	23	24	25	26	27	28	29	30
31	32	33	34	35	36	37	38	39	40
41	42	43	44	45	46	47	48	49	50
51	52	53	54	55	56	57	58	59	60
61	62	63	64	65	66	67	68	69	70
71	72	73	74	75	76	77	78	79	80
81	82	83	84	85	86	87	88	89	90
91	92	93	94	95	96	97	98	99	100

Guided Practice

Use Workmat 5 and a counter.
Subtract.

Think!
Put a counter on 89.
Move up 2 rows.

1. 89 − 20 = ______
2. 35 − 10 = ______
3. 77 − 40 = ______
4. 95 − 30 = ______
5. 43 − 30 = ______
6. 70 − 60 = ______
7. 88 − 20 = ______
8. 64 − 50 = ______

9. 39 − 20 = ______
10. 62 − 50 = ______
11. 56 − 50 = ______
12. 93 − 50 = ______
13. 58 − 40 = ______

14. **Math Talk** Look at 61, 51, 41, and 31 on the hundred chart. Describe the pattern. What number comes next?

Practice

Remember! Move up one row to subtract a ten.

Use the hundred chart or Workmat 5.
Subtract.

15. 31 − 10 = 21

16. 65 − 40 = ______

17. 74 − 30 = ______

18. 89 − 70 = ______

19. 55 − 20 = ______

20. 29 − 10 = ______

1	2	3	4	5	6	7	8	9	10
11	12	13	14	15	16	17	18	19	20
21	22	23	24	25	26	27	28	29	30
31	32	33	34	35	36	37	38	39	40
41	42	43	44	45	46	47	48	49	50
51	52	53	54	55	56	57	58	59	60
61	62	63	64	65	66	67	68	69	70
71	72	73	74	75	76	77	78	79	80
81	82	83	84	85	86	87	88	89	90
91	92	93	94	95	96	97	98	99	100

21. $44 - 20$

22. $67 - 40$

23. $92 - 60$

24. $17 - 10$

25. $90 - 80$

26. $63 - 40$

27. $31 - 20$

28. $86 - 50$

29. $79 - 30$

30. $100 - 20$

Algebra Readiness: Number Sentences

31. Circle the number sentence that is not correct.
Why is it wrong?

59 + 20 = 79   59 − 20 = 39

20 + 59 = 79   20 − 59 = 39

Draw or write to explain.

At Home Ask your child to show you how to find 94 − 50 on the hundred chart.

En casa Pida a su niño que le muestre cómo hallar 94 − 50 en la tabla de números hasta el cien.

Name ______________________________

Chapter 12 Lesson 2

Subtract Using a Place-Value Chart

Learn

Hands On

Objective
Subtract a one-digit number from a two-digit number with and without regrouping.

Linda has 43 bug postcards.
If she gives her sister 5 postcards, how many will she have left?

Find 43 − 5.

Step 1

Show 43.

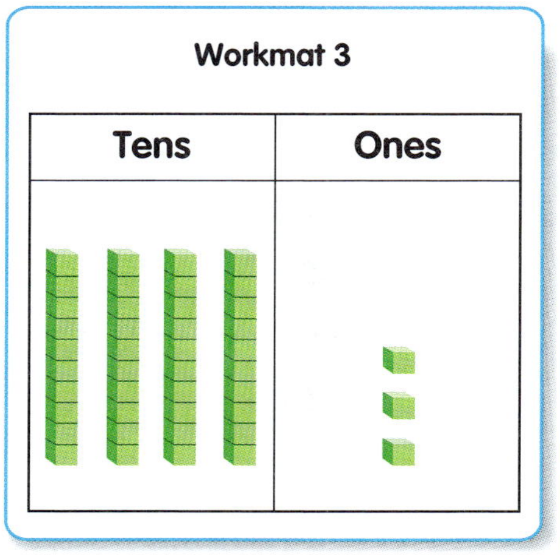

Tens	Ones
☐	☐
4	3
−	5

Step 2

Regroup 1 ten as 10 ones.

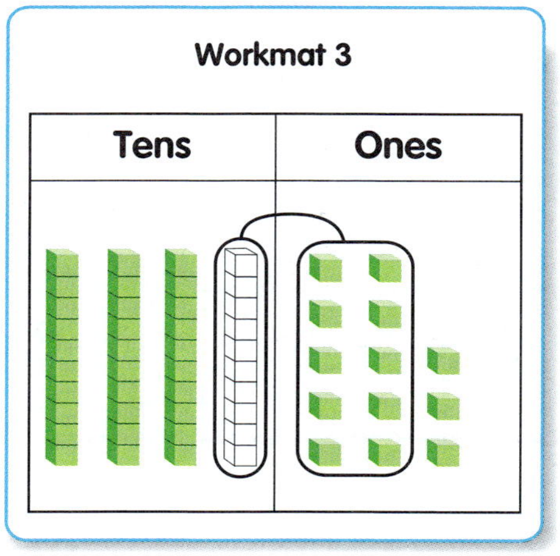

Tens	Ones
3	13
~~4~~	~~3~~
−	5

Step 3

Subtract the ones.
Subtract the tens.

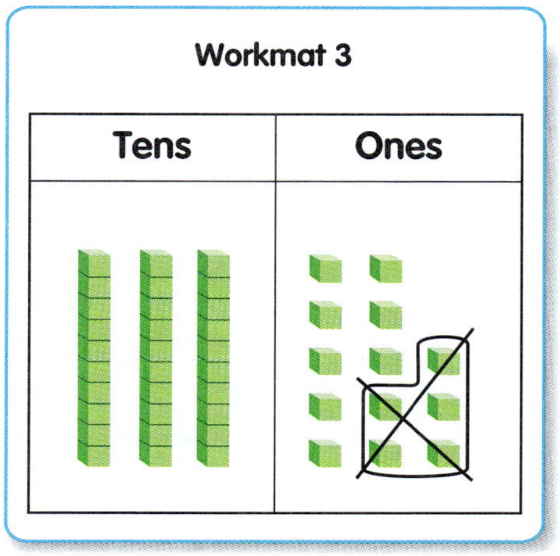

Tens	Ones
3	13
~~4~~	~~3~~
−	5
3	8

Linda has ______ cards left.

Guided Practice

Use Workmat 3 with [tens block] and [ones block].
Subtract. Record each step.

1.

	Tens	Ones
	☐	☐
	3	2
−		4

Think!
Do I need to regroup to subtract 4 ones from 2 ones?

2.

	Tens	Ones
	☐	☐
	5	0
−		6

3.

	Tens	Ones
	☐	☐
	4	6
−		3

4. **Math Talk** What does the 10 you wrote in the ☐ represent in Exercise 2?

Practice

Use Workmat 3 with [tens block] and [ones block].
Record the subtraction.

Remember!
Regroup when there are not enough ones.

5.

	Tens	Ones
	4	10
	~~5~~	~~0~~
−		4
	4	6

6.

	Tens	Ones
	☐	☐
	6	5
−		8

7.

	Tens	Ones
	☐	☐
	8	8
−		3

8.

	Tens	Ones
	☐	☐
	4	5
−		6

Problem Solving: Mental Math

9. Marita finds 34 leaves. She gives some leaves to the caterpillars. Now Marita has 5 leaves left. How many leaves does Marita give away?

_____ leaves

Draw or write to explain.

At Home Tell your child your age. Have your child find the difference between your age and his or her age.

En casa Diga a su niño cuántos años tiene usted. Pida a su niño que halle la diferencia entre la edad de usted y la de él.

Name ______________________________

Objective
Practice subtraction with regrouping.

Chapter 12 Lesson 3

Practice Regrouping with 10 to 14

Learn

Sometimes when you subtract, you regroup to make more ones.

Gabe has 40 bug pencils in his collection. He gives 13 pencils to his brother. How many pencils does Gabe have left?

__27__ pencils

	Tens	Ones
	☐	☐
	4	0
−	1	3

Regroup 1 ten as 10 ones.

	Tens	Ones
	3	10
	~~4~~	~~0~~
−	1	3
	2	7

Guided Practice

Subtract. Regroup if you need to.

1.

	Tens	Ones
	☐	☐
	5	0
−	3	5

Think! Can I subtract 5 ones or do I need to regroup?

2.

	Tens	Ones
	☐	☐
	4	1
−	1	6

3.

	Tens	Ones
	☐	☐
	5	3
−	3	6

4.

	Tens	Ones
	☐	☐
	5	2
−	2	3

5.

	Tens	Ones
	☐	☐
	3	2
−	1	7

6.

	Tens	Ones
	☐	☐
	7	0
−	5	4

7.

	Tens	Ones
	☐	☐
	8	1
−	1	3

8. **Math Talk** What numbers can you subtract from 21 without regrouping? How do you know?

KEY NS 2.2 Find the sum or difference of two whole numbers up to three digits long.
Also AF 1.0, MR 2.0, SDAP 1.3, SDAP 1.4, KEY SDAP 1.0

NS 2.0 Estimate, calculate, and solve problems involving addition and subtraction of two- and three-digit numbers.

Practice

Subtract.
Regroup if you need to.

> **Remember!**
> After you regroup, rewrite the number of tens.

9.

	Tens	Ones
	2	14
	~~3~~	~~4~~
−	1	5
	1	9

10.

	Tens	Ones
	☐	☐
	6	2
−	4	4

11.

	Tens	Ones
	☐	☐
	4	3
−	1	1

12.

	Tens	Ones
	☐	☐
	3	1
−	2	6

13.

	Tens	Ones
	☐	☐
	4	0
−	3	2

14.

	Tens	Ones
	☐	☐
	5	4
−	3	5

15.

	Tens	Ones
	☐	☐
	6	3
−	2	7

16.

	Tens	Ones
	☐	☐
	5	0
−	1	9

Problem Solving: Reasoning

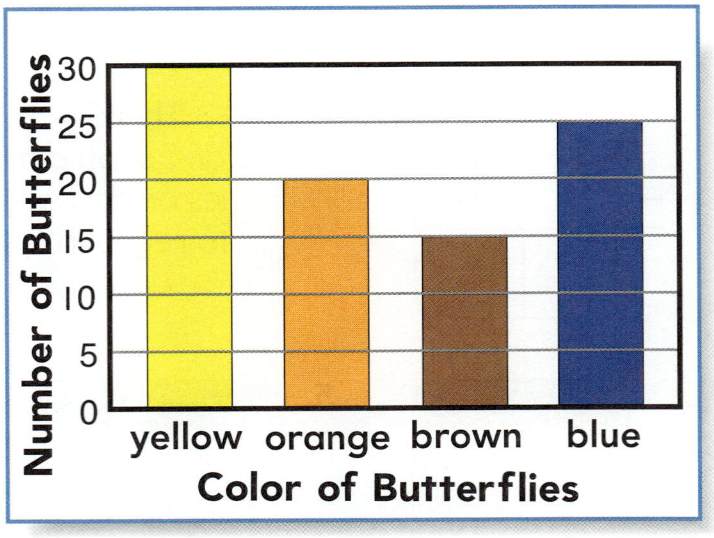

17. Which color butterfly was seen most often?

____

18. How many more blue butterflies were seen than brown butterflies?

______ more blue butterflies

19. Write a subtraction problem using the information in the graph. Then solve.

____

____

At Home Ask your child to solve number sentences such as 60 – 27 and 51 – 34. Ask your child how he or she found the answer.

En casa Pida a su niño que resuelva enunciados numéricos como 60 – 27 y 51 – 34. Anime a su niño a explicar cómo halló la respuesta.

Name ____________________

Chapter 12 Lesson 4

Practice Regrouping with 15 to 18

Objective
Practice subtraction with regrouping.

Learn

Sometimes when you subtract, you regroup to make 15, 16, 17, or 18 ones.

Brandon counts 36 caterpillars in a tree. 19 caterpillars are green. How many caterpillars are not green?

	Tens	Ones
	□	□
	3	6
−	1	9

Regroup 1 ten as 10 ones.

	Tens	Ones
	2	16
	~~3~~	~~6~~
−	1	9
	1	7

17 caterpillars are not green.

Guided Practice

Subtract. Regroup if you need to.

1.

	Tens	Ones
	□	□
	8	7
−	1	8

Think!
Can I subtract 8 ones from 7 ones?

2.

	Tens	Ones
	□	□
	7	5
−	3	9

3.

	Tens	Ones
	□	□
	6	5
−	2	8

4.

	Tens	Ones
	□	□
	7	6
−	1	7

5.

	Tens	Ones
	□	□
	9	6
−	4	9

6.

	Tens	Ones
	□	□
	9	8
−	2	2

7.

	Tens	Ones
	□	□
	5	6
−	1	8

8. **Math Talk** Will you ever need to regroup to make 19 ones? Why or why not?

KEY NS 2.2 Find the sum or difference of two whole numbers up to three digits long. Also **AF 1.3, AF 1.0**

NS 2.0 Estimate, calculate, and solve problems involving addition and subtraction of two- and three-digit numbers.

Practice

Subtract. Regroup if you need to.

Remember!
Always subtract the bottom number from the top number.

9.

	Tens	Ones
	4	17
	~~5~~	~~7~~
−	2	8
	2	9

10.

	Tens	Ones
	4	6
−	2	8

11.

	Tens	Ones
	3	8
−	1	9

12.

	Tens	Ones
	6	6
−	2	9

13.

	Tens	Ones
	2	5
−	1	7

14.

	Tens	Ones
	7	8
−	3	9

15.

	Tens	Ones
	3	6
−	2	5

16.

	Tens	Ones
	4	5
−	1	8

17.

	Tens	Ones
	7	7
−	5	9

18.

	Tens	Ones
	8	6
−	1	8

19.

	Tens	Ones
	5	8
−	3	9

20.

	Tens	Ones
	6	7
−	4	8

Algebra Readiness: Number Sentences

21. Someone spilled juice on the scientist's records. Fill in the numbers that can no longer be read.

Insects Seen	Number of Insects		
	Saturday	Sunday	Total
Dragonflies	____	19	83
Bees	73	21	____
Grasshoppers	67	____	94
Butterflies	71	11	____

At Home Give your child two 2-digit numbers. Ask him or her to subtract the lesser number from the greater. Did your child regroup?

En casa Dé a su niño dos números de 2 dígitos. Pídale que reste el número menor del mayor. ¿Reagrupó su niño?

Name ______________________

Chapter 12 Lesson 5

Reasonable Answers

Problem Solving
Plan

Objective
Decide whether an answer is reasonable.

Learn

Mia has 53 stamps in her collection.
She gives 24 stamps to Jeremy.
She gives some more stamps to Annie.
How many does Mia have left?

15 stamps   30 stamps   45 stamps

Understand

What do you know?

- Mia has 53 stamps.
- She gives 24 stamps to Jeremy.
- She gives some more stamps to Annie.

What do you need to find out?

how many stamps Mia has left

Plan

You can subtract to help find a reasonable answer.

$$\begin{array}{r} 53 \\ -\ 24 \\ \hline 29 \end{array}$$

Mia has 29 stamps left after she gives 24 to Jeremy.

Think!
I can subtract 53 - 24 to find out how many stamps Mia has after she gives Jeremy some.

Solve

Circle the most reasonable answer.

15 stamps   30 stamps   45 stamps

Think!
If Mia gives Annie some of her 29 stamps, I know she cannot have more than 29 stamps left.

Look Back

Is your answer reasonable?
How do you know?

KEY NS 2.2 Find the sum or difference of two whole numbers up to three digits long.
Also AF 1.0, NS 2.0, MR 2.0, MR 2.2

NS 1.0 Understand the relationship between numbers, quantities, and place value in whole numbers up to 1,000.

Guided Problem Solving

1. Caleb has 48 cups in his collection. Alex has 27 cups in his collection. Danny has more cups than Alex but fewer than Caleb. How many cups does Danny have?

20 cups   35 cups   60 cups

What do you need to find out?

____

Think!
I have to pick the number that is between 27 and 48.

Circle the most reasonable answer.

20 cups   35 cups   60 cups

2. **Math Talk** How did you choose a reasonable answer?

Problem Solving Practice

Circle the most reasonable answer.

Draw or write to explain.

3. Lucia has 61 crayons. She gives 37 crayons to her brother. She also gives some crayons to her sister. How many crayons does Lucia have left?

10 crayons   30 crayons   50 crayons

4. Nathan has 35 stickers. Beth has 24 stickers. Tom has fewer stickers than Beth. How many stickers does Tom have?

40 stickers   30 stickers   20 stickers

At Home Ask your child to explain how he or she chose a reasonable answer in Exercise 3.

En casa Pida a su niño que explique cómo eligió una respuesta razonable para el Ejercicio 3.

Name ______________________________

California Field Trip

At the Shipley Nature Center

You can see many kinds of birds, such as brown pelicans, great blue herons, mallards, and California quail. At Blackbird Pond, you can see willow trees. There is also a butterfly garden.

Blackbird Pond

Choose a way to solve. Show your work.

1. There are 51 mallards swimming in Blackbird Pond. Some mallards fly out of the pond. How many mallards are left?

   70 mallards   55 mallards   35 mallards

mallard

2. Volunteers at the nature center count 49 downy woodpeckers in one year. In the second year, they count 16 fewer downy woodpeckers. How many woodpeckers do they count in the second year?

   ______ woodpeckers

downy woodpecker

3. Laura counts 37 monarch butterflies. Carlos counts 25 monarch butterflies. How many more butterflies does Laura count?

   ______ more butterflies

monarch butterfly

KEY NS 2.2 Find the sum or difference of two whole numbers up to three digits long.
Also AF 1.0, NS 2.0, MR 1.0, MR 1.1
Science LS 2.a

Problem Solving on Tests
Listening Skills

Listen to your teacher read the problem.
Choose the correct answer.

Select a Strategy
Act It Out
Write a Number Sentence
Reasonable Answers

1. Allie has 50¢. She has 1 quarter, 2 dimes, and one other coin. What is the fourth coin?

dime	quarter	nickel	penny
○	○	○	○

NS 5.0

2. Luca has 35 butterfly pictures. Shannon has 7 butterfly pictures. How many butterfly pictures do they have in all?

32	42	52	62
○	○	○	○

KEY NS 2.2

3.

92	82	72	62
○	○	○	○

KEY NS 2.2

4.

about 50	about 60	about 70	about 80
○	○	○	○

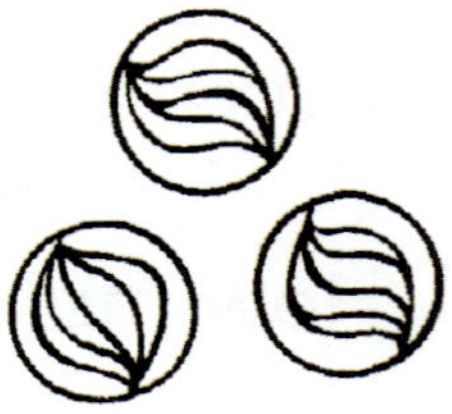

NS 6.0

MR 1.0 Make decisions about how to set up a problem.
MR 1.1 Determine the approach, materials, and strategies to be used.

Education Place
Visit **www.eduplace.com/camap/** for Test-Taking Tips and Extra Practice.

Name ______________________

Key Standards Review

Find the difference.

1. 95 − 30 = ______
2. 72 − 20 = ______
3. 89 − 43 = ______
4. 53 − 11 = ______
5. 86 − 53 = ______
6. 74 − 23 = ______
7. 73 − 50 = ______
8. 58 − 26 = ______
9. 44 − 13 = ______

Challenge Number Sense

This is what a checkerboard looks like before you begin a game. Use the game board to answer the questions.

1. How many spaces are on the board in all?

   ______ spaces

2. How many checkers are on spaces on the board?

   ______ checkers

3. Write a number sentence to show how many spaces do not have checkers on them.

   ______ − ______ = ______ spaces

KEY NS 2.2 Find the sum or difference of two whole numbers up to three digits long.

Real World Data

California Trees

Marta visits Muir Woods in Marin County, California. The fog makes this forest damp and cool all year round. The tall redwood trees provide shade for the shorter trees.

Heights of Trees

California Buckeye	30 feet
Red Alder	60 feet
Coast Live Oak	50 feet
California Bay Laurel	80 feet

Marta makes a table to show how tall some trees can grow. She wants to show the results as both a picture graph and a bar graph.

1. Use the data in the table to make a picture graph.
2. Use the data in the table to make a bar graph.
3. Use your graphs. Write two word problems. Trade problems with a classmate and solve.

Heights of Trees

California Buckeye	
Red Alder	
Coast Live Oak	
California Bay Laurel	

Key: Each ● = 10 feet

Heights of Trees

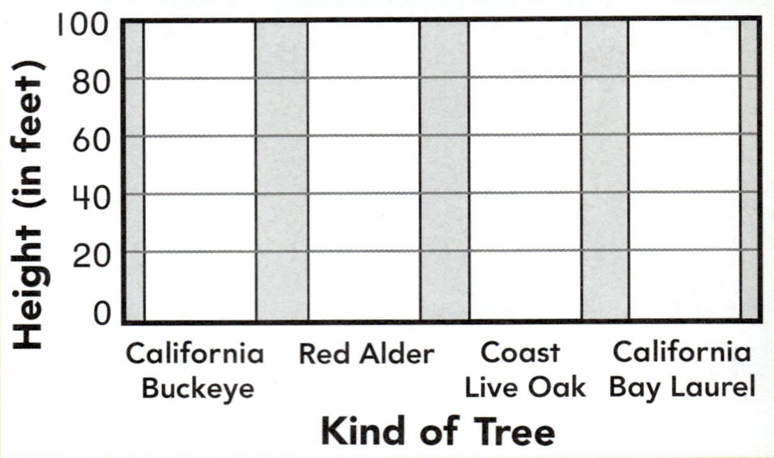

KEY SDAP 1.0 Collect numerical data and record, organize, display, and interpret the data on bar graphs and other representations.
Also **SDAP 1.2**
Science LS 2.e, IE 4.e

Chapter 12 Review/Test

Name ______________________

Concepts and Skills

Use the hundred chart or Workmat 5.
Subtract. KEY NS 2.2

1	2	3	4	5	6	7	8	9	10
11	12	13	14	15	16	17	18	19	20
21	22	23	24	25	26	27	28	29	30
31	32	33	34	35	36	37	38	39	40
41	42	43	44	45	46	47	48	49	50
51	52	53	54	55	56	57	58	59	60
61	62	63	64	65	66	67	68	69	70
71	72	73	74	75	76	77	78	79	80
81	82	83	84	85	86	87	88	89	90
91	92	93	94	95	96	97	98	99	100

1. 65 − 30 = ______

2. 82 − 70 = ______

Subtract. Regroup if you need to. KEY NS 2.2

3.

	Tens	Ones
	☐	☐
	5	3
−		5

4.

	Tens	Ones
	☐	☐
	5	4
−	1	8

5.

	Tens	Ones
	☐	☐
	8	2
−		7

6.

	Tens	Ones
	☐	☐
	7	5
−	2	9

Problem Solving KEY NS 2.2, MR 2.0

7. There is an ocean fish exhibit at the aquarium. There were 20 blue fin tuna and 15 hammerhead sharks. Then, they released 18 fish back to the ocean. How many fish are left in the exhibit?

   ______ fish

   Explain how you know.

   ____

   ____

Draw or write to explain.

Spiral Review and Test Practice

1. Which is 77 rounded to the nearest ten?

70 ○ 75 ○ 77 ○ 80 ○

NS 6.0 Page 71

2. Which is the value of the coins?

60¢ ○ 75¢ ○ $1.00 ○ $1.60 ○

KEY NS 5.2, KEY NS 5.1, NS 5.0 Page 127

3. What is the sum?

44¢ + 9¢ = ______

449¢ ○ 44¢ ○ 49¢ ○ 53¢ ○

NS 5.0, KEY NS 5.2, KEY NS 2.2, NS 2.0 Page 195

4. Which is the difference?

$$\begin{array}{r} 42 \\ -\ \ 7 \\ \hline \end{array}$$

49 ○ 45 ○ 40 ○ 35 ○

KEY NS 2.2 Page 233

Education Place
Visit www.eduplace.com/camap/ for Test-Taking Tips and Extra Practice.

Chapter 13

Relate 2-Digit Addition and Subtraction

Vocabulary

Here are some vocabulary words you will learn in the chapter.

round To find the nearest ten

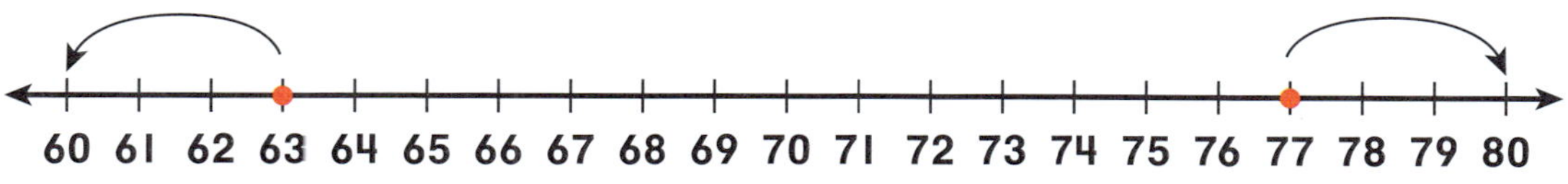

63 is closer to 60.
63 rounds down to 60.

77 is closer to 80.
77 rounds up to 80.

estimate A rounded answer that is close to an exact answer

48 —round to nearest ten→ 50
−31 —round to nearest ten→ −30
20 ← estimate

check subtraction You can check subtraction with addition.

$$\begin{array}{r} 74 \\ -25 \\ \hline 49 \end{array} \qquad \begin{array}{r} 49 \\ +25 \\ \hline 74 \end{array}$$

See English-Spanish Glossary pages 573–589.

KEY NS 2.2 Find the sum or difference of two whole numbers up to three digits long.
Also KEY NS 2.1, NS 6.0

Education Place
Visit www.eduplace.com/camap/ for the eGlossary and eGames.

Name ______________________

Check What You Know

Solve. Use the number line.

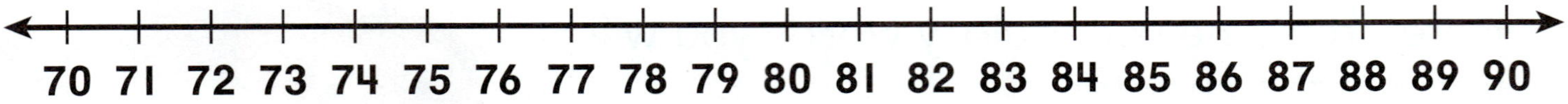

1. Does 86 round to 80 or 90?

   86 rounds to ______.

2. Use the nearest tens to estimate the difference of 86 – 73.

   ______ – ______ = ______

3. Find the exact difference of 86 – 73.

   ______ – ______ = ______

4. Does 85 round to 80 or 90?

   85 rounds to ______.

5. Use the nearest tens to estimate the difference of 85 – 75.

   ______ – ______ = ______

   Find the exact difference of 85 – 75.

   ______ – ______ = ______

Use this page to review important skills needed for this chapter.

Name ______________________

Chapter 13 Lesson 1

Different Ways to Subtract

Hands On

Objective
Choose an appropriate computation method.

Learn

Different ways to subtract:
- mental math
- hundred chart
- tens and ones blocks
- paper and pencil

Choose a way to find
55 − 39.

Guided Practice

Circle the way you would use to subtract.
Find the difference.

Think!
Can I count back in my head to subtract 10 from 36?

1. 36 − 10 = _____
   - mental math
   - paper and pencil

2. 41 − 32 = _____
   - tens and ones blocks
   - mental math

3. 87 − 43 = _____
   - paper and pencil
   - tens and ones blocks

4. 58 − 30 = _____
   - mental math
   - hundred chart

5. 96 − 37 = _____
   - mental math
   - tens and ones blocks

Explain how you find the difference.

6. 47 − 13 ____

____

7. **Math Talk** Why might it be faster to use mental math than paper and pencil to find 73 − 30?

Practice

Circle the way you would use to subtract.
Find the difference.

Remember!
Choose the way that works best for you.
- mental math
- hundred chart
- tens and ones
- paper and pencil

8. 72 − 15 = 57

mental math

paper and pencil

9. 67 − 17 = ____

hundred chart

mental math

10. 92 − 58 = ____

paper and pencil

mental math

11. 57 − 22 = ____

tens and ones blocks

paper and pencil

12. 41 − 39 = ____

hundred chart

mental math

Subtract. Explain how you find the difference.

13. 42 − 19

14. 58 − 7

15. 84 − 37

16. 88 − 79

Problem Solving: Number Sense

17. Billy has 80 dinosaur stickers. He gives 30 away. Sonya asks Billy for 60 stickers. Does he have enough to give her? Explain how you know.

Draw or write to explain.

At Home Have your child show different ways to find 58 − 20 and explain which way he or she would use.

En casa Pida a su niño que muestre diferentes maneras de hallar 58 − 20 y que explique qué manera usaría.

Name ______________________________

Chapter 13 Lesson 2

Estimate Differences

Objective
Estimate differences by rounding to the nearest ten.

Vocabulary
round

Learn

When you do not need an exact answer, you can estimate.

Estimate the difference of 35 – 22.

Step 1

Round each number to the nearest ten.

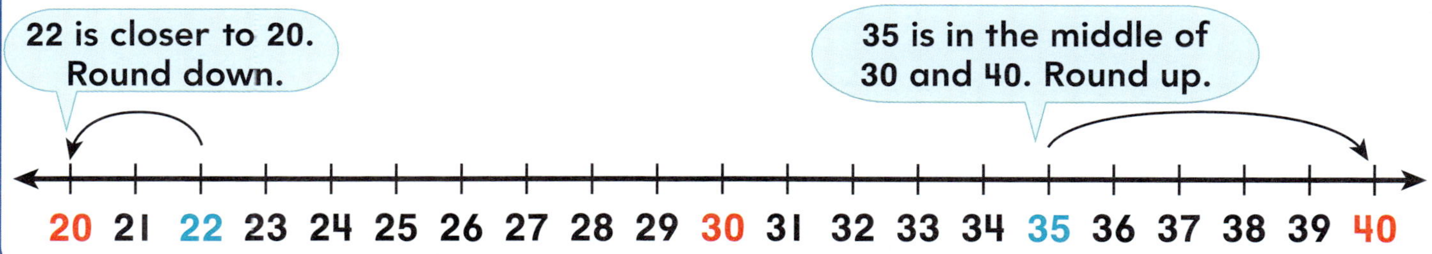

Step 2

Subtract the tens to estimate the difference.

40 – 20 = 20

Guided Practice

Round each number to the nearest ten.
Estimate the difference.

Think!
Both numbers have 5 or more ones, so I round up.

1. 49 – 35

   _____ – _____ = _____

2. 41 – 32

   _____ – _____ = _____

3. 45 – 28

   _____ – _____ = _____

4. **Math Talk** How did you find the nearest ten for each number in Exercise 3?

NS 6.0 Use estimation strategies in computation and problem solving that involve numbers that use the ones, tens, hundreds, and thousands places. Also **MR 2.0**

NS 2.0 Estimate, calculate, and solve problems involving addition and subtraction of two- and three-digit numbers.

Practice

Round each number to the nearest ten.
Estimate the difference.

5. 58 – 42

60 – 40 = 20

6. 53 – 41

____ – ____ = ____

7. 49 – 43

____ – ____ = ____

8. 57 – 34

____ – ____ = ____

9. 59 – 47

____ – ____ = ____

10. 44 – 41

____ – ____ = ____

11. 45 – 18

____ – ____ = ____

12. 37 – 21

____ – ____ = ____

13. 49 – 33

____ – ____ = ____

14. 46 – 24

____ – ____ = ____

15. Estimate 56 – 24.

____ – ____ = ____

Find the exact difference.

56 – 24 = ____

Problem Solving: Reasoning

16. Karen estimates 57 – 11 this way:

50 – 10 = 40

Is her estimate reasonable? Explain.

Draw or write to explain.

At Home Ask your child to demonstrate how to round to the nearest ten and estimate to find 42 – 23.

En casa Pida a su niño que demuestre cómo redondear a la decena más cercana y cómo estimar para hallar 42 – 23.

Name ____________________

Chapter 13 Lesson 3

Line Up Digits to Subtract

Objective
Rewrite subtraction exercises in vertical form.

Learn

Find 85 – 59.

Step 1

Rewrite 85 – 59.
Line up the ones and tens.

	Tens	Ones
	☐	☐
	8	5
–	5	9

Step 2

Subtract the ones.
Subtract the tens.

	Tens	Ones
	7	15
	~~8~~	~~5~~
–	5	9
	2	6

Guided Practice

Write the numbers in vertical form. Subtract.

1. 46 – 7

Think!
Do I write the 7 in the ones or the tens column?

2. 63 – 7

3. 84 – 55

4. 81 – 32

5. 77 – 58

6. 78 – 69

7. 65 – 4

	Tens	Ones
	☐	☐
–		

8. **Math Talk** Brenda rewrites 72 – 6. She subtracts and gets a difference of 12. What did she do wrong?

Practice

Write the numbers in vertical form. Subtract.

9. 36 – 9

	Tens	Ones
	2	16
	~~3~~	~~6~~
–		9
	2	7

10. 58 – 24

	Tens	Ones
	☐	☐
–		

11. 68 – 32

	Tens	Ones
	☐	☐
–		

Remember!
Line up the ones
and the tens.

12. 64 – 47

	Tens	Ones
	☐	☐
–		

13. 42 – 3

	Tens	Ones
	☐	☐
–		

14. 71 – 25

	Tens	Ones
	☐	☐
–		

15. 38 – 17

	Tens	Ones
	☐	☐
–		

16. 96 – 34

	Tens	Ones
	☐	☐
–		

17. 81 – 4

	Tens	Ones
	☐	☐
–		

18. 60 – 35

	Tens	Ones
	☐	☐
–		

19. 45 – 27

	Tens	Ones
	☐	☐
–		

Problem Solving: Reasoning

20. Tanya has 45 beads. She uses 17 to make a necklace. She uses 13 to make a bracelet. How many beads does Tanya have left?

_____ beads

Draw or write to explain.

At Home Have your child show you how to rewrite 53 – 16 and 63 – 7 and then find the differences.

En casa Pida a su niño que muestre cómo escribir 53 – 16 y 63 – 7 de otra manera y que halle las diferencias.

Name ______________________________

Chapter 13 Lesson 4

Add and Subtract Money

Objective
Add and subtract money amounts.

Learn

You can add and subtract money like whole numbers. Just remember to write the ¢ sign.

Cameron wants to buy 2 ladybug postcards. How much do 2 ladybug postcards cost?

$$\begin{array}{r} 15 \\ +15 \\ \hline 30 \end{array} \qquad \begin{array}{r} 15¢ \\ +15¢ \\ \hline 30¢ \end{array}$$

If Cameron pays with 50¢, how much change does he get back?

$$\begin{array}{r} 50 \\ -30 \\ \hline 20 \end{array} \qquad \begin{array}{r} 50¢ \\ -30¢ \\ \hline 20¢ \end{array}$$

Guided Practice

Add or subtract.

1. $\begin{array}{r} 54¢ \\ -16¢ \\ \hline \end{array}$

Think! 54¢ – 16¢ is like 54 – 16.

2. $\begin{array}{r} 19¢ \\ +\ 7¢ \\ \hline \end{array}$

3. $\begin{array}{r} 71¢ \\ -37¢ \\ \hline \end{array}$

4. $\begin{array}{r} 41¢ \\ -32¢ \\ \hline \end{array}$

Rewrite the numbers.
Then add or subtract.

5. 39¢ + 43¢

	3	9¢
+	4	3¢

6. 94¢ – 6¢

–		

7. 50¢ – 13¢

–		

8. 51¢ + 19¢

+		

9. **Math Talk** How is adding money similar to adding numbers?

KEY NS 5.2 Know and use the decimal notation and the dollar and cent symbols for money. Also **NS 2.0, NS 5.0, MR 3.0, MR 2.0**

KEY NS 2.2 Find the sum or difference of two whole numbers up to three digits long.

Practice

Add or subtract.

Remember!
Write the ¢ sign in your answer.

10. 80¢	11. 25¢	12. 33¢	13. 72¢	14. 65¢
−37¢	+46¢	+50¢	− 6¢	−45¢
43¢				
15. 58¢	16. 44¢	17. 55¢	18. 34¢	19. 60¢
−39¢	+ 8¢	+27¢	−16¢	−37¢

Solve.

20. Adriana has 8 dimes. She buys an album and a bag of marbles. How much money does she have left?

_____¢

21. If Adriana buys 2 albums instead, how much money does she have left?

_____¢

Problem Solving: Reasoning

22. Cindy has 7 dimes. She buys two marbles. How much money does she have left?

_____¢

Draw or write to explain.

At Home Give your child a few coins. Ask him or her if there is enough money to buy the bag of marbles shown above.

En casa Dé a su niño algunas monedas. Pregúntele si hay suficiente dinero para comprar la bolsa de canicas que se muestra arriba.

Name ____________________

Chapter 13 Lesson 5

Check Subtraction

Objective
Check subtraction by using addition.

Learn

You can add to check your subtraction.

Step 1

Subtract.

	Tens	Ones
	4	16
	~~5~~	~~6~~
−	1	8
	3	8

Step 2

Start with the difference.
Add the number you subtracted.

	Tens	Ones
	1	
	3	8
+	1	8
	5	6

If the sum equals the number you subtracted from, your answer is correct.

Guided Practice

Subtract. Check by adding.

1. 32 − 19 = ___ Check: ___ + ___ = ___

Think!
13 is the difference, so I can add 13 and 19 to check.

2. 81 − 66 = ___ Check: ___ + ___ = ___

3. 72 − 64 = ___ Check: ___ + ___ = ___

4. **Math Talk** Why can you use addition to check subtraction?

Practice

Subtract. Then add to check.

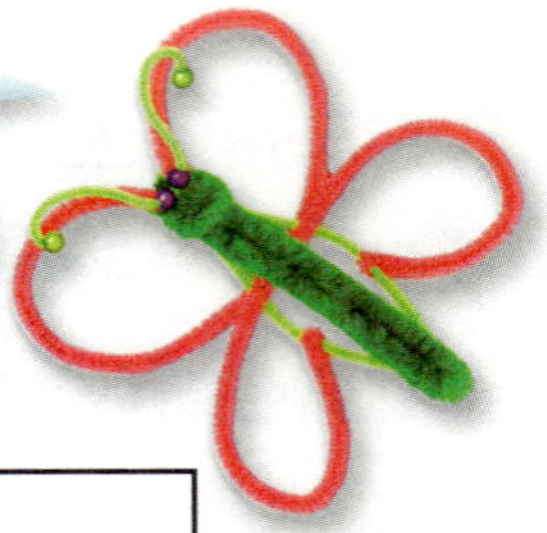

Remember! Add the difference and the number you subtracted to check.

5. $\begin{array}{r} 68 \\ -\ 42 \\ \hline 26 \end{array}$ $\begin{array}{r} 26 \\ +\ 42 \\ \hline 68 \end{array}$

6. $\begin{array}{r} 40 \\ -\ 5 \\ \hline \end{array}$ $\begin{array}{r} \square \\ +\ \square \\ \hline \square \end{array}$

7. $\begin{array}{r} 54 \\ -\ 19 \\ \hline \end{array}$ $\begin{array}{r} \square \\ +\ \square \\ \hline \square \end{array}$

8. $\begin{array}{r} 85 \\ -\ 6 \\ \hline \end{array}$ $\begin{array}{r} \square \\ +\ \square \\ \hline \square \end{array}$

9. $\begin{array}{r} 49 \\ -\ 28 \\ \hline \end{array}$ $\begin{array}{r} \square \\ +\ \square \\ \hline \square \end{array}$

10. $\begin{array}{r} 74 \\ -\ 10 \\ \hline \end{array}$ $\begin{array}{r} \square \\ +\ \square \\ \hline \square \end{array}$

11. 53 bees are buzzing in the meadow. 7 bees fly away. How many bees are left?

$\begin{array}{r} \square \\ -\ \square \\ \hline \square \end{array}$ $\begin{array}{r} \square \\ +\ \square \\ \hline \square \end{array}$

12. 98 bees are in the hive. 51 bees fly out. How many bees are left in the hive?

$\begin{array}{r} \square \\ -\ \square \\ \hline \square \end{array}$ $\begin{array}{r} \square \\ +\ \square \\ \hline \square \end{array}$

Algebra Readiness: Number Sentences

13. 43 + ▲ = 82

What is ▲? ______

14. 96 − ■ = 14

What is ■? ______

At Home Ask your child to show you how to use addition to check a subtraction problem such as 58 – 29.

En casa Pida a su niño que le muestre cómo usar la suma para comprobar un problema de resta como 58 – 29.

Name ____

Chapter 13 Lesson 6

Create and Solve

Problem Solving

Objective
Write problems and use strategies to solve them.

The table shows the prices for packs of bird cards.

Card Prices	
Super Pack	80¢
Big Pack	60¢
Extra Pack	35¢
Basic Pack	22¢

1. How much more does the Super Pack cost than the Extra Pack?

   The Super Pack costs _____¢ more.

2. Write your own subtraction problem about the bird cards. Use the prices in the table.

   ____

   ____

   ____

   ____

   ____

3. Write the subtraction sentence. Line up the digits to subtract. Solve.

   −
   _____

Share your subtraction problem with a classmate.

Problem Solving

The table shows the prices for toy insects.

Toy Insect Prices	
Insect	**Price**
Ladybug	49¢
Grasshopper	35¢
Butterfly	26¢
Cricket	42¢
Ant	18¢

4. Melissa wants to buy a ladybug and a cricket. How much will she spend in all?

   Melissa will spend _____¢.

5. Write your own addition problem about the toy insects. Use the prices in the table.

   ____

   ____

   ____

   ____

   ____

6. Write the addition sentence. Line up the digits to add. Solve.

   \+ ____

Share your addition problem with a classmate.

Name ______________________________

Key Standards Review

Find the difference.

1. 75 − 37 = ______
2. 85 − 39 = ______
3. 53 − 14 = ______
4. 45 − 28 = ______
5. 92 − 56 = ______
6. 73 − 66 = ______
7. 81 − 65 = ______
8. 53 − 39 = ______
9. 84 − 69 = ______

Challenge — Number Sense

Write 6, 7, 8, and 9 in the first set of boxes to get the greatest possible difference.

Write 6, 7, 8, and 9 in the second set of boxes to get the least possible difference.

1. **Greatest**

```
   □ □
 − □ □
 ─────
```

2. **Least**

```
   □ □
 − □ □
 ─────
```

KEY NS 2.2 Find the sum or difference of two whole numbers up to three digits long.

How Many Will Be Left?

Math Music, Track 5
Tune: original

Let's subtract two numbers.
What's one thing you might do?
If you don't have enough ones—
Regroup! It might help you.
Regroup 1 ten as ten ones—
To see if this helps you!

Juan collects 26 stamps
And gives 19 to me.
How many stamps will he have left?
Regroup and we will see!

Sue collects 32 bugs
And gives 15 to me.
How many bugs will she have left?
Regroup and we will see!

Keb collects 44 rocks
And gives 16 to me.
How many rocks will he have left?
Regroup and we will see!

KEY NS 2.2 Find the sum or difference of two whole numbers up to three digits long.

Name ______________________________

Concepts and Skills

Subtract.
Explain how you found the difference. KEY NS 2.2, MR 2.0

1. 56 − 4 = _____

____

____

Round each number to the nearest ten. Estimate the difference. NS 6.0

2. 78 − 24

_____ − _____ = _____

3. 45 − 33

_____ − _____ = _____

Write the numbers in vertical form. Subtract. KEY NS 2.2

4. 87 − 29

Tens	Ones
□	□
−	

Add or subtract. KEY NS 2.2

5.
$$\begin{array}{r} 56¢ \\ +\ 15¢ \\ \hline \end{array}$$

6.
$$\begin{array}{r} 72¢ \\ -\ 34¢ \\ \hline \end{array}$$

Subtract. Then add to check. KEY NS 2.1

7.
$$\begin{array}{r} 94 \\ -\ 37 \\ \hline \end{array}$$
□ + □

8.
$$\begin{array}{r} 91 \\ -\ 8 \\ \hline \end{array}$$
□ + □

Problem Solving KEY NS 5.1, MR 2.0

9. Rhonda has 3 quarters. She buys a toy horse for 56¢. How much money does Rhonda have left? _____

Spiral Review and Test Practice

1. If Pam pays for the item below with these coins, what is her change?

4¢	6¢	8¢	10¢
○	○	○	○

KEY NS 5.1, KEY NS 5.2, NS 5.0 Page 129

2. What is the sum?

$$\begin{array}{r} 5 \\ 23 \\ +56 \\ \hline \end{array}$$

84	79	74	28
○	○	○	○

KEY AF 1.1, AF 1.0, KEY NS 2.2, NS 2.0 Page 197

3. Which is the difference?

$$\begin{array}{r} 84 \\ -15 \\ \hline \end{array}$$

71	79	69	61
○	○	○	○

KEY NS 2.2 Page 233

4. Round each number to the nearest ten. Which is an estimate for the difference? $78 - 12$

90	80	70	60
○	○	○	○

NS 6.0, KEY NS 2.2, NS 2.0 Page 249

Education Place
Visit **www.eduplace.com/camap/** for Test-Taking Tips and Extra Practice.

Name ______________________________

Greg Tang's **Go Fast, Go Far**

Unit 5 Mental Math Strategies

Partial Differences

Partial differences are quick. Subtracting pieces is the trick!

I break the numbers into pieces. Then I subtract the pieces. 30 − 20 = 10 and 8 − 4 = 4. The answer is 10 + 4 = 14.

1. 38 − 24 = [10] + [4] = [14]
2. 45 − 23 = [20] + [ ] = [ ]
3. 67 − 32 = [ ] + [ ] = [ ]
4. 89 − 45 = [ ] + [ ] = [ ]
5. 26 − 12 = [ ] + [ ] = [ ]
6. 35 − 23 = [ ] + [ ] = [ ]
7. 47 − 15 = [ ] + [ ] = [ ]
8. 58 − 34 = [ ] + [ ] = [ ]

Take It Further: Now try all the steps in your head!

9. 36 − 22 = [ ]
10. 98 − 31 = [ ]

Reading and Writing Math

The table below shows how many stamps Robbie has. How many more beetle stamps than butterfly stamps does he have?

Kind of Stamp	Number of Stamps
Beetle	35
Butterfly	28

Word Bank
difference
regroup
subtract
ten

Choose a word from the word bank for each blank.

1. To solve the problem, you ________ 35 − 28.

2. When you ________ in subtraction, you are trading 1 ________ for 10 ones.

3. The result of a subtraction problem is called the ________.

4. **Writing Math** Write a subtraction problem and tell which way you used to solve it. Explain why you chose that way.

____

____

____

KEY NS 2.2 Find the sum or difference of two whole numbers up to three digits long.
Also AF 1.2, MR 1.1

Name ____________________

Concepts and Skills

Complete the subtraction sentences.
Use quick pictures or a basic fact to help. KEY NS 2.2

1. 9 tens − 4 tens = _____ tens

_____ − _____ = _____

Use the hundred chart or Workmat 5.
Subtract. KEY NS 2.2

2. 64 − 20 = _____

3. 85 − 40 = _____

1	2	3	4	5	6	7	8	9	10
11	12	13	14	15	16	17	18	19	20
21	22	23	24	25	26	27	28	29	30
31	32	33	34	35	36	37	38	39	40
41	42	43	44	45	46	47	48	49	50
51	52	53	54	55	56	57	58	59	60
61	62	63	64	65	66	67	68	69	70
71	72	73	74	75	76	77	78	79	80
81	82	83	84	85	86	87	88	89	90
91	92	93	94	95	96	97	98	99	100

Record the subtraction. KEY NS 2.2

4.

	Tens	Ones
	☐	☐
	7	6
−		9

5.

	Tens	Ones
	☐	☐
	4	0
−		8

Subtract. Regroup if you need to. KEY NS 2.2

6. $\begin{array}{r} 52 \\ -\ \ 6 \\ \hline \end{array}$

7. $\begin{array}{r} 75 \\ -38 \\ \hline \end{array}$

8. $\begin{array}{r} 64 \\ -16 \\ \hline \end{array}$

Subtract. Explain how you found the difference. KEY NS 2.2, MR 2.1

9. 66 − 4 = ______

How I solved: ______________________________

Round each number to the nearest ten.
Estimate the difference. NS 6.0

10. 98 − 30

______ − ______ = ______

98 − 30 is about ______.

11. 57 − 32

______ − ______ = ______

57 − 32 is about ______.

Add or subtract. KEY NS 2.2

12. 76¢ + 19¢

13. 68¢ − 39¢

Subtract. Then add to check. KEY NS 2.1

14. 64 − 27

□ + □

Problem Solving

Solve. MR 2.0, KEY NS 2.2

15. Lee-Ann and Paul both have stickers. Paul has 33 stickers. Together they have 70 stickers in all. How many stickers does Lee-Ann have?

______

Draw or write to explain.